EXTRATERRESTRIAL LIFE

EXTRATERRESTRIAL LIFE

are we alone?

ROBERT IRION

The Reader's Digest Association, Inc.
Pleasantville, New York/Montreal

EXTRATERRESTRIAL LIFE

READER'S DIGEST PROJECT STAFF

Editorial Director
David Schiff

Editor
Charles Flowers

Senior Designer
George McKeon

Production Technology Manager
Douglas A. Croll

Editorial Manager
Christine R. Guido

CONTRIBUTORS

Author
Robert Irion

Art Director
Eleanor Kostyk

Copy Editor
Nora Reichard

Technical Reviewer
Frank Drake

Picture Research
Carousel Research, Inc.

Photo Editors
Laurie Platt Winfrey
Mary Teresa Giancoli

Photo Researchers
Van Bucher
Christopher Deegan
Cristian Peña

READER'S DIGEST ILLUSTRATED REFERENCE BOOKS

Editor-in-Chief
Christopher Cavanaugh

Art Director
Joan Mazzeo

Operations Manager
William J. Cassidy

Printed in the United States of America.

To order additional copies of *Out There: Extraterrestrial Life*,
call 1-800-846-2100.

You can also visit us on the World Wide Web at: **www.readersdigest.com**

Library of Congress Cataloging in Publication Data

Irion, Robert.
Extraterrestrial life: are we alone?/Robert Irion.
p. cm.—(Out there)
ISBN 0-7621-0318-3
1. Life on other planets—Popular works. I. Title. II. Series.

QB54 .I75 2001
576.8'39—dc21 00-067326

Address any comments about *Out There: Extraterrestrial Life* to:
Reader's Digest
Editor-in-Chief, Illustrated Reference Books
Reader's Digest Road
Pleasantville, NY 10570

The more I examine the universe and study the details of its architecture, the more evidence I find that the universe in some sense must have known we were coming.

—Freeman Dyson

Disturbing the Universe

EXTRA-TERRESTRIAL LIFE

are we alone?

OUR MOST PROFOUND QUESTION

Planet Earth teems with life. It inhabits every nook and pore, every rocky crevice and watery droplet. Plants and animals thrive on the land, in the air, and in the sea, making Earth a blue-green oasis in the apparent desert of the universe. Is that oasis unique? Or are there others like it, rich with life that we may or may not recognize?

In the 4 billion years or so since life first appeared on Earth, only one species has evolved enough to

As the twenty-first century begins, we know of only one place in the universe that harbors life: our planet Earth.

This Norwegian alpine scene is, as far as we know, unique to our solar system.

ponder the question of whether we are alone. Humans have long debated the answer, but science has provided little evidence one way or the other. Now that situation is changing. Scientists may soon discover how life arose from a rich stew of organic matter in the turmoil of our planet's youth. They are learning how primitive organisms have adapted to live in settings from scalding deep-sea vents to harsh fields of ice, often far from any source of sunlight. They are realizing that such life-supporting conditions probably exist elsewhere in our galaxy, perhaps even on a handful of planets or moons in our solar system.

Further, researchers are planning missions that will search for signs of life on other worlds. Some missions will probe for living things on Mars or on Jupiter's moon Europa, which probably has an icy ocean. Others will look for Earth-like planets circling stars within 200 light-years of our Sun, scanning for imprints of life in their atmospheres. And an ongoing project, the Search for Extraterrestrial Intelligence, or SETI, uses radio and optical telescopes to hunt for intentional or accidental signals from any advanced civilizations that may have arisen in our Milky Way Galaxy.

The emerging science combining all of these disciplines is called astrobiology. Many astrobiologists think that at least one of their searches will show that we have companions in the cosmos. That would be a profound moment, if only because many people deeply believe that we occupy a privileged position in the hierarchy of the universe. Even if we find the equivalent of pond scum rather than intelligent life, we will still have a new awareness that life has arisen elsewhere. However, astrobiologists may face a steep challenge to find an entity far from Earth that is, beyond all doubt, alive.

A bacterium divides in half, one of the hallmarks of a living cell. These *Escherichia coli* bacteria, magnified fifty thousand times, are common residents of our intestines.

THE ORIGIN OF LIFE

What does it mean to be alive? Scientists must keep some definition in mind as they try to detect life beyond Earth. Otherwise, subtle evidence of life elsewhere might escape their notice. But as obvious as it may seem, the line between "life" and "nonlife" is not easily drawn.

The fact is, living things and most inanimate objects are composed of the same raw materials. A few common elements—mostly hydrogen, carbon, oxygen, and nitrogen—and compounds made from those elements are the key ingredients of life, but they exist in many nonliving things as well. It's easy to look at a rock and a tree and tell which one is alive. But on the scale of bacteria and amoebas, it's not so clear.

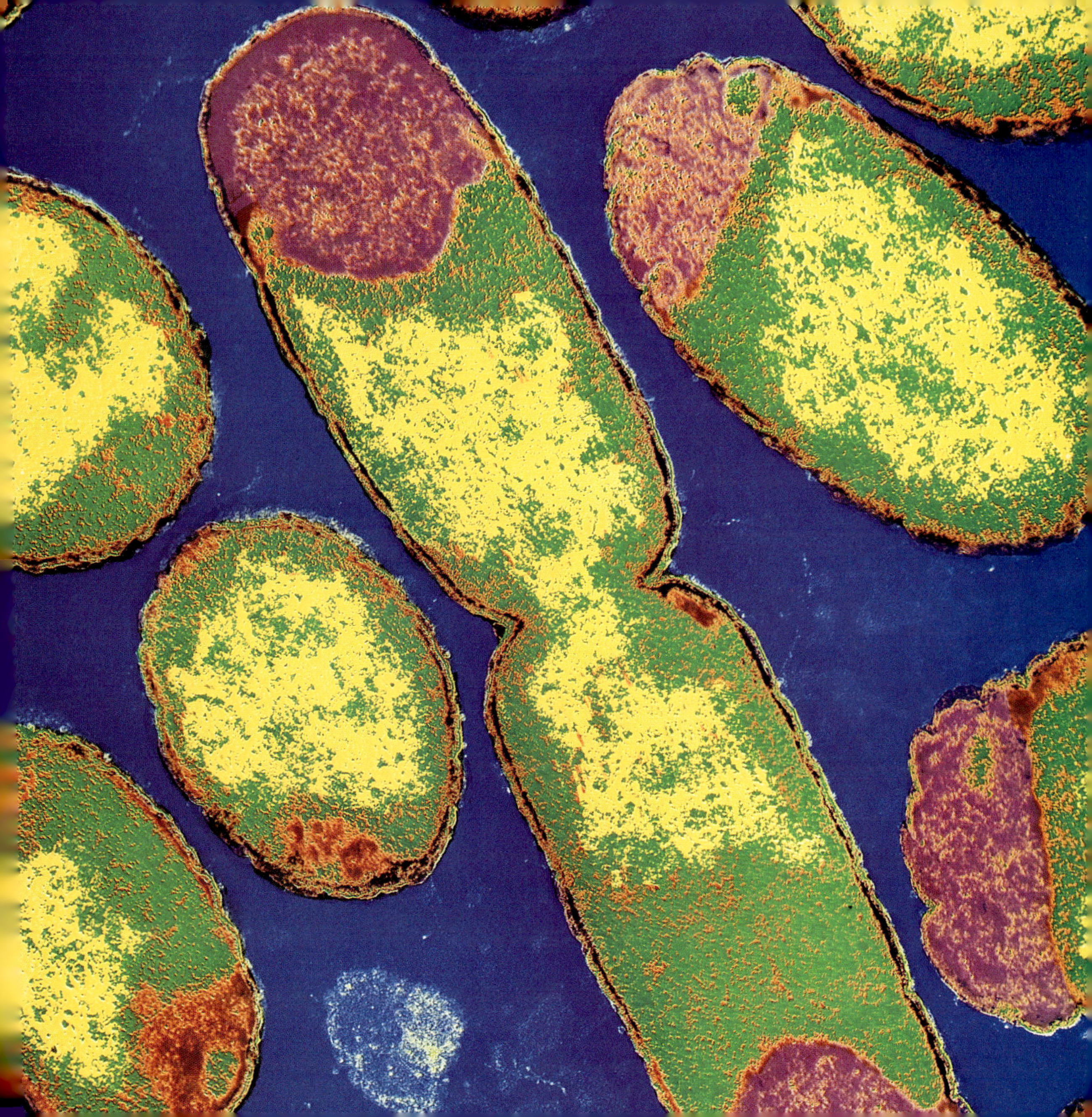

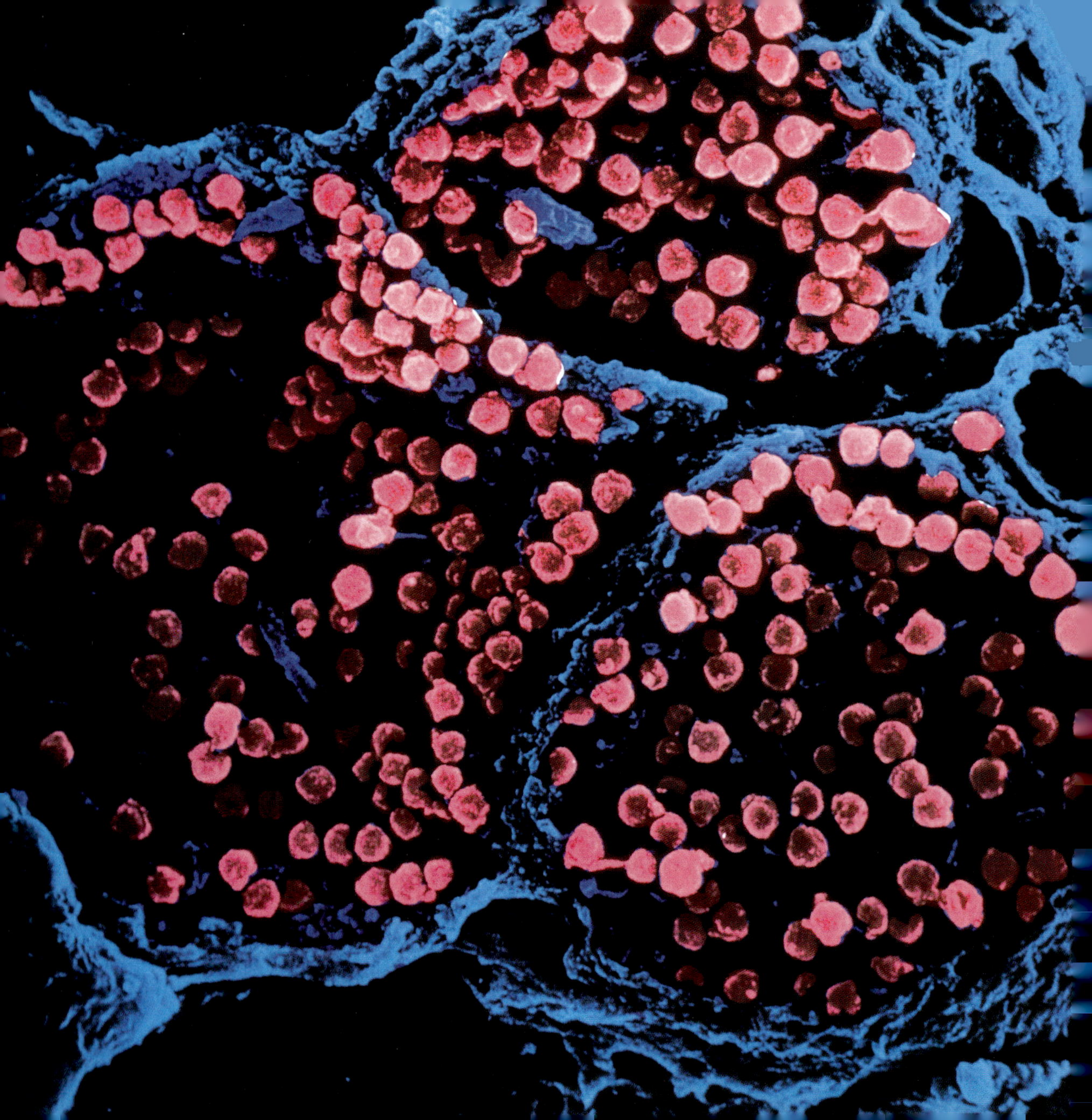

Baculovirus particles (pink) infect the cells of an insect in this electron micrograph photo. A virus must co-opt the genetic machinery of a host cell in order to reproduce.

One signpost of life as we know it is a boundary that separates the living thing from its surroundings, such as the membrane of a cell. Within that boundary, life sustains itself in an orderly way. That counteracts entropy, the tendency in nature for everything to become more disordered with time. Yet this definition is not sufficient by itself. Salt crystals and computer monitors also keep order within their boundaries, but they are not alive. Therefore, scientists add that a living thing must create its own boundary by using fuel from its environment. This process, called metabolism, also gives an organism the energy it needs to function.

Still, something more fundamental distinguishes cheetahs from cars. Life must grow, and it must reproduce, perpetuating itself via its progeny. That act can be as simple as a bacterium dividing in half or as complex as a human embryo gestating inside its mother. Most biologists go further to say that life must also evolve, mutating and adapting to changes in its environment. Offspring that inherit beneficial changes are more likely to survive and pass them along to the next generation. Charles Darwin first described that critical process of natural selection more than a century ago. It's an important concept in astrobiology as well, for scientists assume that organisms elsewhere also must cope with changing conditions.

Armed with these guidelines, biologists feel confident that they can identify life on Earth. Even so, there are gray areas. For instance, is a virus alive? A single virus is a simple and inert particle. In order to make copies of itself, it must invade a host's cells and hijack each cell's reproductive machinery. Viruses may not meet all the definitions of life, but they contain life's basic genetic material in tiny, efficient packages. They may be widespread in the universe, so scientists will take measures to ensure that no harmful viruses or other infectious agents return to Earth from other worlds.

The next step for astrobiologists is to use Earth as a model for examining how life arose. Those origins are shrouded by more than 4 billion years of time. As far as we know, no traces remain of the first organisms. Even so, careful studies of today's primitive cells combined with biochemical experiments in the lab have provided clues.

It's immediately apparent that life on Earth requires organic molecules, which have carbon as a primary ingredient. One of the most common elements in the universe, carbon combines easily with itself and with other elements to form a wide array of stable compounds. Life as we know it also requires liquid water, which helps organic materials react easily and bathes the fragile building blocks of life in a protective coating.

This computer model depicts a molecule of DNA, a basic genetic component of life on Earth. Colors represent the atoms of five elements: hydrogen (white), carbon (gray), oxygen (red), nitrogen (blue), and phosphorus (orange).

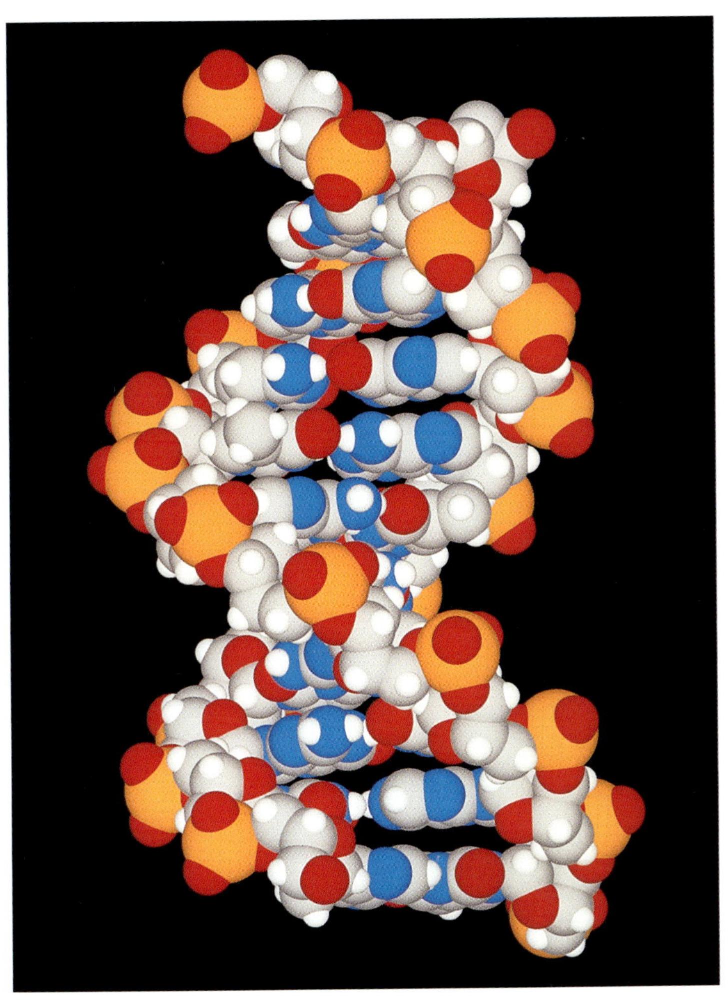

Impacts by comets may have delivered some of the building blocks of life to the early Earth. In this composite photo, Comet Hale-Bopp hangs above eerie mineral deposits, called tufa towers, in Mono Lake, California.

Scientists think that life's essential compounds, such as sugars and amino acids, had two main sources early in Earth's history. First, lightning from storms and ultraviolet light from the Sun probably sparked organic compounds to form out of gases in the planet's atmosphere, creating a steady rain of them onto the surface. Second, astronomers have learned that asteroids and comets are rich in organic substances. Those objects bombarded Earth for hundreds of millions of years, possibly enriching our young planet with the literal seeds of life. It is also possible that some vital compounds have bubbled forth at hot spots on the ocean floor.

Tide pools, such as this quiet refuge in Brazil's Fernando de Noronha islands, may have served as cradles for the first living cells.

The first living cells may have coalesced in a calm, watery environment, such as a tide pool at the seashore. With warmth from the Sun and steady nourishment from the ocean, such a setting seems ideal for the organic broth of early Earth to simmer. On the other hand, some scientists observe that the most primitive organisms on Earth today live in hot places, such as geothermal pools or deep-sea volcanic vents. Life may have arisen in one of these places first, they speculate, spurred by the chemical reactions of carbon-rich compounds at high temperatures.

In either case, two big mysteries remain. No one knows how or why the first membranes enclosed some of the organic slurry into

These rocks near Almeria, Spain, may look like ordinary boulders but they actually are stromatolites—the fossilized remains of ancient colonies of blue-green algae that grew in mound-shaped layers. The oldest stromatolites are preserved in Australian rocks as old as 3.5 billion years, among the earliest signs of life on earth.

self-contained cells. It may have happened by chance. When it did, those cells enjoyed a major advantage, for their interiors were shielded from the rigors of open water. But for that advantage to mean anything, the cells also had to copy themselves. That's the second and more profound mystery: How did the first reproduction occur? Today's living cells use the sophisticated interplay of two genetic molecules, DNA and RNA, to forge the proteins that carry out life's basic functions. The molecules also preserve genetic information and transmit it from one generation to the next. The initial copying mechanisms must have been simpler, but biologists suspect that some version of proteins and RNA—which has a simpler chemical structure than DNA—were involved.

This basic blueprint of origins may be followed elsewhere, especially where liquid water is abundant, but the details might change from one world to the next. For instance, extraterrestrial life may not use the same genetic molecules found here to reproduce. Cells may evolve a different way of tapping their star's light for energy, or they may draw sustenance from chemical compounds within their planet. It's conceivable that carbon is not the only foundation for life; sulfur is one possible alternative. Perhaps silicon-based cells swarm across the surface of a distant planet, like animated computer chips.

Life on Earth has evolved to thrive not only on land but also in the air (water bat, right), and in the sea (oceanic whitetip shark, below).

Modern life often echoes Earth's past. Ancestors of today's ferns arose on land about 400 million years ago (top). Extinct trilobites (middle) ruled the seafloor for hundreds of millions of years. Their descendants include the primitive-looking horseshoe crab (bottom).

One conclusion is clear: Each life-form will have adapted to the temperature, gravity, and natural resources unique to its own world. Finding such alien beings will require not only a careful search but a vivid imagination as well.

Heat-loving bacteria thrive within the Morning Glory geothermal spring, near the Old Faithful geyser in Yellowstone National Park, Wyoming.

LIFE'S SURPRISES ON EARTH

The more scientists learn about life on Earth, the more they appreciate how resilient and adaptable it is. That's good news for the prospects of extraterrestrial life—at least, life of the single-celled variety. The case may not be as convincing for more complex organisms.

The first single-celled life probably arose about 4 billion years ago, within a few hundred million years of Earth's birth. The oldest fossils that scientists have identified are layered mats of bacteria, called stromatolites, preserved in 3.5-billion-year-old rocks. For eons, Earth was the realm of these primitive cells. They spread and diversified into every conceivable niche on the planet, evolving to survive under seemingly hostile conditions.

For example, some bacteria and algae inhabit scalding water in the geothermal springs of Yellowstone National Park and other hot spots around the world. Not only do they live there, they thrive: they reproduce most efficiently at temperatures near the boiling point of

3lind crabs and ɔther bizarre animals survive on ɔacteria at hydrothermal vents, such as Saracen's Head n the mid-Atlantic Ocean, two miles deep. Hot sulfur-rich fluids spew from the vent, heated by volcanic activity below.

water. Even more extraordinary are bacterial communities that cluster near superheated water and minerals expelled from thermal vents on the ocean floor, more than a mile deep. These organisms draw upon hydrogen sulfide as their energy source, since no sunlight penetrates to those depths. Tube worms, clams, and other creatures feed on the bacteria, resulting in bizarre ecosystems unlike anything we see at the surface.

At the other end of the scale, many microbes have evolved to tolerate extreme cold. Bacteria eke out a minimal existence in tiny pockets of brine encased within ice in the Arctic and Antarctic, eating and reproducing infrequently. Some cells appear to secrete substances that lower the freezing point of the water around them, thus keeping their habitats liquid. In other cases, ice acts as a preservative. Ancient cells can remain dormant within ice for thousands or millions of years, reawakening when temperatures grow warm enough to melt their chambers.

Hot and cold are not the only extremes to which organisms on Earth have adapted. Some bacteria live in fluids that are more corrosive than battery acid. Others grow slowly within rocks or far underground, tapping minerals for the energy they need. Still others withstand intense pressures, dehydration, and doses of radiation that would quickly destroy the cells in our bodies.

These various organisms, which scientists call extremophiles, have much to teach us about extraterrestrial life. Their presence in so many different settings on Earth suggests that it wouldn't take much to get life started on other worlds. However, that's a far cry from saying that complex life is common elsewhere. Indeed, it took at least 2 billion years after life arose on Earth for evolution to create the first multicellular life-forms, while plants and animals appeared only within the last 400 million to 700 million years. Does this history mean that the path toward advanced life on other worlds might be just as capricious?

That question is provoking a vigorous debate among astrobiologists today. Some argue that many astronomical and geological circumstances have worked together to make Earth uniquely hospitable to higher life-forms. That combination of circumstances, they maintain, would be so rare on other worlds that we shouldn't expect an interstellar telegram anytime soon.

For instance, our Sun is a long-lived star, with an output of energy that has not varied markedly for billions of years. It orbits in a part of the Milky Way Galaxy that is rich with carbon, oxygen, and other elements essential for life, supplied by the explosions of past generations of giant stars. In addition, the Sun is not too close to

Our Sun is an average star with a stable lifetime of 10 billion years —more than enough time for life to evolve and thrive on Earth.

Volcanic eruptions, such as this plume from the Tolbachik volcano on Russia's Kamchatka peninsula, supply gases to Earth's life-friendly atmosphere.

the center of the galaxy, where supernovas and near misses with other stars would wreak havoc.

Within our solar system, Jupiter acts as a "protector planet." Its powerful gravitational influence propels many dangerous comets and asteroids into deep space, away from possible collision courses with Earth. By stabilizing the tilt of Earth's axis, our own Moon also has helped life evolve on Earth. Without that steadying influence, temperatures might have fluctuated too wildly for complex life to gain a toehold.

Earth itself has fostered the growth of its biological riches. The planet has just the right composition for its outer crust to break up into relatively thin slabs and move relentlessly. Those motions, called plate tectonics, help to regulate the gases in our atmosphere via volcanic activity. They also may stimulate new species by rearranging the positions of the continents. Earth's molten core generates a magnetic field that traps most of the dangerous high-energy particles from the Sun, preventing them from striking the surface. Our atmosphere is also just right. If it was much thicker, the planet might boil over into an inferno like Venus. If it was much thinner, Earth might freeze solid like Mars.

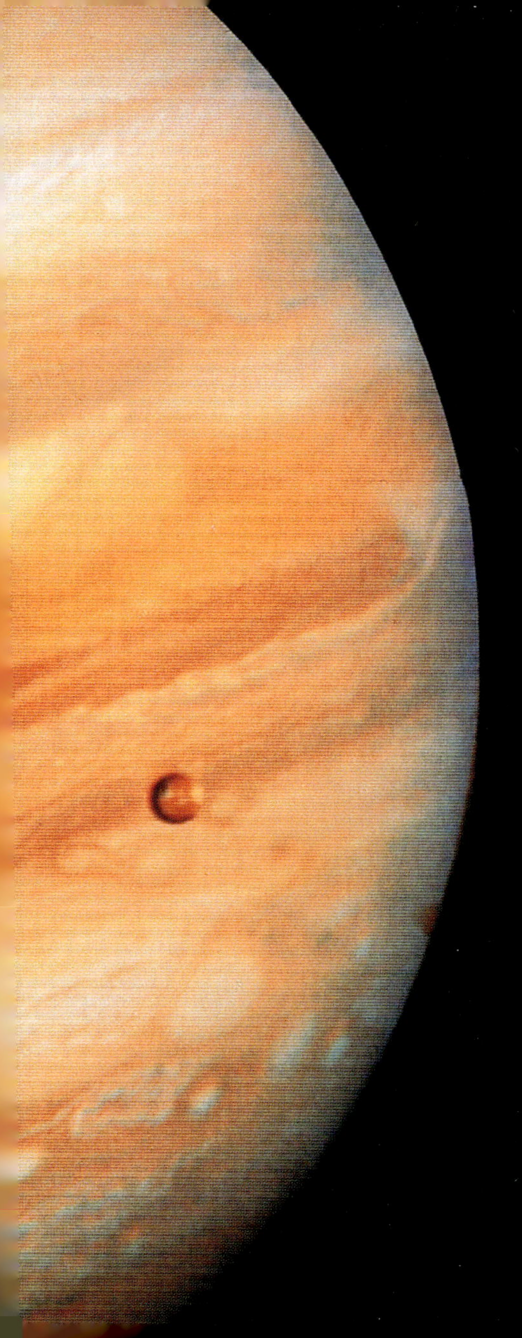

Jupiter, the king of planets, fosters life on Earth by deflecting comets and asteroids that might otherwise strike our planet. Shown in this *Voyager 1* photo are two of Jupiter's largest moons: Io (mid-right in front of planet) and Europa (above).

Indeed, everything about Earth seems eerily perfect for plants and animals to evolve and thrive. Might those conditions be unique in our galaxy? Many scientists think not. They counter partly with a statistical argument. The Milky Way contains at least 100 billion stars. Many nearby stars have planets, so the galaxy must have billions of them. The raw ingredients of life are common in space, and evolution seems bound to occur on worlds that have stable surfaces with liquid water. It's impossible to know how often those conditions are met. Still, the claim that our planet is the sole abode for advanced life strikes these researchers as far-fetched—and perhaps as egocentric as the old notion that everything revolved around Earth.

Further, they point to the dazzling variety of life on Earth today as evidence that complex organisms are as hardy and adaptable as microbes. At several times since multicellular life has arisen, catastrophes have threatened to wipe out life on Earth entirely. The most famous of these was the asteroid collision 65 million years ago that generated huge forest fires and a globe-circling cloud of dust, probably eradicating half the species on Earth, including the dinosaurs. But in each case, life rebounded, and a new mix of species became dominant. It seems that once higher life-forms appear on the cosmic stage, it's not easy to lower the curtain upon them.

This bald eagle, bringing fish to chicks in its nest on Alaska's Aleutian Islands, is emblematic of the rich variety of advanced life on Earth today.

Intelligent life has evolved to take many forms on our planet, including these playful bottle-nosed dolphins. However, our species remains the only one capable of searching for life that may exist on other worlds.

LIFE IN OUR SOLAR SYSTEM?

Although no one expects that any other planet or moon in our solar system hosts a biological cornucopia equal to Earth's, planetary scientists now believe that microbial life, at least, might find conditions to its liking on other worlds within the Sun's family. The leading contenders are Mars, our planetary neighbor, and Europa, one of Jupiter's large moons.

Robotic probes started exploring our solar system in earnest in the mid-1970s. Two *Viking* landers analyzed soil on Mars in 1976. (In the inset, the landers are shown at a Colorado test site.) Meanwhile *Voyager 2* flew past all four giant outer planets—including Neptune (bottom) and its largest moon, Triton, in 1989.

Scientific attitudes about Mars have come full circle during the last one hundred years. Turn-of-the-century astronomer Percival Lowell thought he perceived canals and changing vegetation patterns on the Red Planet from his observatory in Flagstaff, Arizona. Studies with bigger telescopes cast doubt upon those illusions. Then, the *Viking* mission to Mars dashed them in 1976. An orbiting spacecraft photographed a dry and dusty planet, while two landers sampled soils that apparently were barren of life. If organisms had arisen on Mars, they seemed ancient history.

By contrast, findings from recent missions have invoked the ghost of Lowell, but with a few twists. More detailed satellite images from a mission called Mars Global Surveyor have revealed networks of valleys and canyons that could have been carved only by running water. Water also appears to have burst from the steep walls of some

Water may have streamed from the walls of a crater in the Noachis Terra region of Mars millions of years ago, according to this stunning photo from the Mars Global Surveyor space probe. The same orbiter found an extremely smooth region around the Martian north pole (inset, in blue), the possible imprint of an ancient ocean.

craters on Mars. In addition, the satellite images show a huge expanse of land around the Martian north pole that is far flatter than any land on Earth. Scientists speculate that an ocean covered this land in the past.

The surface of Mars is dry and cold today, and any liquid water would instantly evaporate into space because of the low atmospheric pressure. However, the polar caps contain frozen water and carbon dioxide (dry ice), and liquid water may exist within underground layers warmed by the planet's internal heat. If life arose on Mars billions of years ago, when the surface of the planet may have been wet, it probably migrated to the depths along with the water. Scientists plan to send future landers and rovers to areas on Mars where water has flowed to search for elusive evidence that life once thrived there.

Europa may present an even more intriguing case. This moon is the second of Jupiter's four Galilean satellites, so named because the Italian scientist Galileo Galilei discovered them with his small

Europa, an ice-covered moon of Jupiter, probably has an ocean beneath its cracked outer shell. The *Galileo* spacecraft captured these views of the moon, including complex ridges and faults between giant rafts of ice (inset).

telescope in 1610. Europa's exterior is a smooth icy shell laced with cracks. It appears that a liquid ocean flows beneath this cold shield, far deeper than oceans on Earth. Europa is too distant for the Sun's warmth to melt its ice. Rather, the heat comes from friction caused by an ongoing tug-of-war among Jupiter, Europa, and neighboring moons. The combined gravitational pulls of these objects exert tides upon Europa that flex its interior to and fro. This action warms its rocks and ices, just as shaking a bag of ice cubes would melt the surfaces of the cubes.

Images of Europa's surface from the *Galileo* spacecraft in the late 1990s convinced scientists that water lurks beneath the ice. Much of the ice is broken up into rafts that look like chunks of frozen ocean on Earth. Also, there is evidence that a layer within Europa influences the patterns of magnetic fields in space around it. The most likely explanation for this phenomenon is a thick, salty ocean.

Prospects for life within this strange sea are tantalizing. Although no light would penetrate into the ocean, thermal energy from the gravitational tides would keep its bottom layers warm. There might even be deep volcanic ridges as in Earth's oceans. Organic matter from comets and space dust would enter the ocean as the surface ice drifts and cracks. Filled with salts and carbon-rich compounds, the

Europan ocean might be a dark equivalent of the primitive organic stew from which life may have arisen on Earth.

Fortuitously, one place on Earth may bear some resemblance to this distant sea. It is Lake Vostok, a freshwater lake the size of Lake Ontario trapped beneath several miles of ice in Antarctica. Scientists believe that Lake Vostok has been isolated for tens of millions of years, yet they see bacterial cells trapped within the ice just above the lake. Planners of a mission to search for life on Europa would like to practice on Lake Vostok, but they must take care that their probes will not foul the pristine lake with surface microbes. A mission to drill into the Europan ocean will be difficult and expensive because of the long journey and the technological challenges. However, it seems only a matter of time until a robot from Earth lowers itself into the waters of an alien world within our solar system.

Dozens of lakes lie buried beneath miles of ice in central Antarctica, including 150-mile-long Lake Vostok. Because these ice-covered lakes are not visible from space, the position of Lake Vostok has been outlined on this satellite photo of Antarctica. Scientists may drill into Lake Vostok to help prepare for a future mission to Jupiter's moon Europa.

OTHER STARS, OTHER WORLDS

Direct sampling of worlds beyond our solar system won't happen for a long time, if ever. The costs of interstellar travel are prohibitive, not to mention the many centuries or millennia it would take to reach even the nearest star system. If astronomers hope to detect life out there, they must devise indirect tests.

LAKE VOSTOK

THE SEARCH FOR EXTRATERRESTRIAL INTELLIGENCE

Are there other civilizations in our galaxy? SETI, the Search for Extraterrestrial Intelligence, seeks to answer that question. Of course, SETI scientists have no clue what an extraterrestrial civilization might look like, and fanciful conceptions such as the one shown here play no role in SETI research. Rather, SETI uses the world's largest radio telescopes to listen for artificial signals that other civilizations—regardless of their appearance—may emit into the Milky Way. Our own society emits signals every day because our radio and TV broadcasts stream into space at the speed of light.

SETI astronomers believe that advanced beings do exist elsewhere. This belief is not based upon blind faith. Instead, they point to a sober analysis of a mathematical equation—a grand calculation of the odds that we are not the highest form of life in the galaxy. Look inside for a summary of this formula in words and pictures.

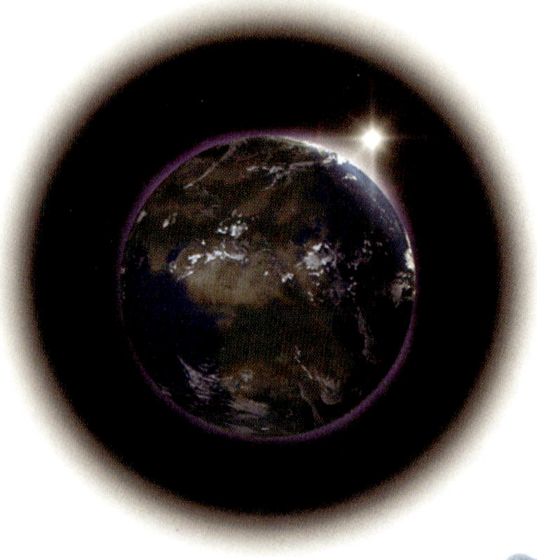

The average number of "Earths" in each of those planetary systems: planets or moons where temperatures allow liquid water (a probable requirement of life) to flow

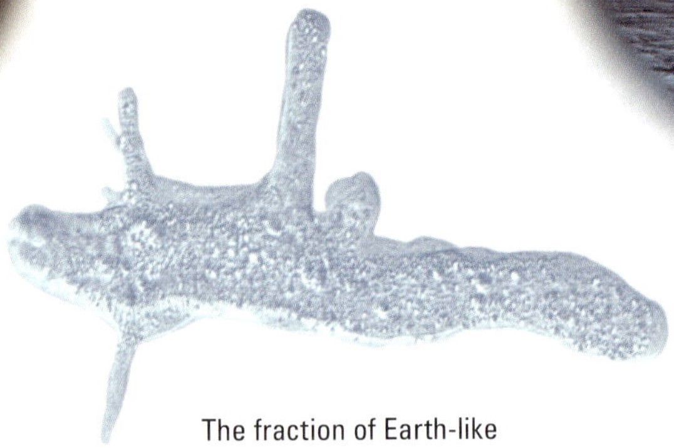

The fraction of Earth-like bodies upon which life of any kind arises

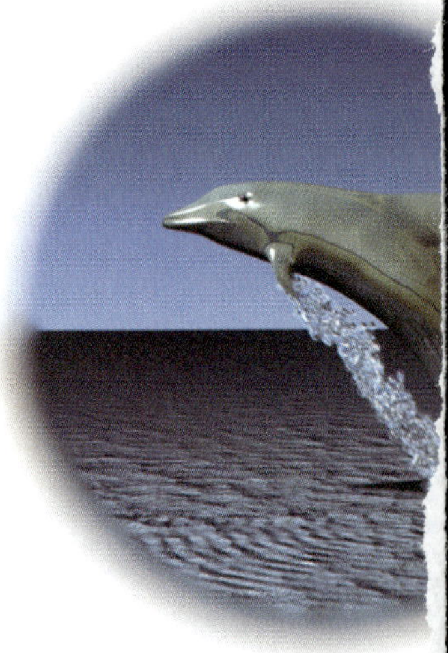

The fraction of life-bea
worlds upon which int
life develops

ring
lligent

The fraction of such worlds where intelligent life evolves into technological civilizations capable of emitting signals that travel deep into space

The average lifetime, in years, before those intelligent civilizations become extinct or stop emitting detectable signals

Our Milky Way, which probably looks similar to the spiral galaxy shown here, may contain tens of billions of planetary systems around other stars.

The first step is to determine how frequently planets and moons form around other stars. Models suggest that while planets are common, planetary systems come in many shapes and sizes. Some may have rocky Earth-like planets at reasonable distances from their stars, while others may be dominated by gaseous planets the size of Jupiter and Saturn. Life is less likely if there are too many giant planets, because the giants fling smaller objects out of planetary systems like rocks from a slingshot.

In the late 1990s, astronomers made dramatic progress toward understanding the population of planets in our Milky Way. They concentrated on stars like our Sun, an average-size star that has a stable life span of about 10 billion years. Their technique was not to look for planets directly—a difficult task for optical telescopes, since dim planets are lost in the blinding glare of their stars. Rather, the astronomers measured the starlight with clever tools that exposed tiny jiggling motions by the stars themselves. The jiggles arise when the gravitational pull of an unseen planet tugs a star back and forth slightly as the planet orbits. That motion stretches and compresses the light waves from the star via the Doppler effect, the same phenomenon that causes a motorcycle to sound high-pitched as it approaches and low-pitched as it recedes.

Analysis of jiggles by dozens of stars revealed some surprising trends. Some giant planets orbit their stars incredibly quickly, in just three or four days. That means they are about ten times closer to their stars than Mercury is to our Sun. Theorists have a hard time explaining how those planets got there and how they survive, but clearly they would be blazingly hot worlds, unsuitable for life. Other giant planets swoop close to their stars and then far away, like comets instead of well-behaved planets in circular orbits. Temperatures would fluctuate too much to support life on such planets or their moons. Further, their erratic motions would disrupt other planets, perhaps scattering them into space.

So far, the planetary systems unveiled by this technique look nothing like our own. However, that's not discouraging. If our own solar system is any indication, rocky planets would be similar in size to Venus or Earth. Such planets would be too small to produce detectable wobbles in their parent stars. Also, among the planets that are large enough to detect, more tend to be the size of Saturn or smaller rather than gigantic like Jupiter. If that trend continues to the scale of rocky planets, the Milky Way could brim with potential habitats for life.

Astronomers envision other ways to spot Earth-like objects for a more accurate planetary census. One method is to watch for the

A Jupiter-size planet orbits around the star Rho Corona Borealis in this artist's conception. The planet is so close to its star that it completes one "year" in just forty Earth days.

The Keck 1 telescope (left) and its twin, Keck 2 (not shown), have spotted most of the known planets beyond our solar system. They sit atop Mauna Kea in Hawaii, near NASA's Infrared Telescope Facility (right).

"shadows" of planets as they cross in front of their stars. Just as a moth fluttering across a street lamp will dim its light slightly, a distant planet will reduce the glow of its star if the planet's orbit happens to cut across our line of sight to the star. The effect is tiny, much less than a 1 percent reduction in brightness. Nevertheless, astronomers have proposed a satellite called Kepler, which would monitor many thousands of stars at once to spot such dimmings. The naming of this satellite recognizes German astronomer Johannes Kepler, who deciphered the orbits of the planets in our solar system.

Another promising way to see Earth-size planets is to use an optical trick called nulling to eliminate the light from their stars. Stars are at least a million times brighter than the reflected light from their planets. However, astronomers have learned how to use two or more mirrors to force pairs of light waves from a star to cancel each other out, like the crest of one wave hitting the trough of another. That leaves behind the faint light from planets for telescopes to discern. This nulling technique works with ground-based telescopes but will be most effective from an orbiting observatory. To that end, NASA plans to launch an ambitious satellite in 2010 called the Terrestrial Planet Finder. It will search for small planets close to their stars—the most likely homes for life in our galaxy.

Future satellites may stare at crowded fields of stars in our Milky Way Galaxy to search for the shadows of companion planets. The star at top center is Sirius, the brightest star in the sky.

It may seem that such planets will lie forever beyond our reach and beyond proof that life has arisen upon them, but reversing our perspective suggests otherwise. With the right equipment, an astronomer in a distant planetary system could conclude without doubt that Earth is swarming with life. The giveaway is our atmosphere. Specifically, the blanket of air around our planet is rich with oxygen. Plants and certain bacteria churn oxygen into the air as they use sunlight to consume carbon dioxide and water for energy. That reaction, called photosynthesis, has profoundly altered Earth's atmosphere during the last 2.5 billion years. Without it, all free oxygen in the air would quickly react with rocks and soils—and higher life-forms could not have evolved.

If the Terrestrial Planet Finder discovers any Earth-like planets, it and other powerful telescopes on the ground will scrutinize the light reflected from them. Oxygen and its reactive cousin, ozone, leave distinctive imprints upon this light. Such signatures would be strong clues that plant life exists on the distant world. On the other hand, if the light reveals an atmosphere rich in carbon dioxide, then the planet probably would resemble cloud-covered Venus or barren Mars. At the very least, we could rule out that photosynthetic life has taken root there.

However, even the most convincing signature of an oxygen-rich planet would say nothing about whether advanced life-forms—and possibly even civilizations—share the planet's surface with plants and bacteria. That's a completely different challenge, and a considerably more provocative one.

THE SEARCH FOR EXTRATERRESTRIAL INTELLIGENCE

Finding microbes on another planet would cause a temporary media frenzy. On the other hand, hearing a greeting from intelligent aliens would have a far more lasting impact. Such a signal would forever alter our sense of the universe and our place in it.

For more than forty years, a few dedicated astronomers have endeavored to catch those greetings, eavesdrop on interstellar chats, or otherwise prove their conviction that our species has company. Their ongoing projects are collectively known as the Search for Extraterrestrial Intelligence (SETI). The U.S. government canceled all federal funding of SETI in 1993, but the research is going strong with private support. Among the general public, it arouses passions like few other topics in astronomy today.

The basic tenet of SETI is that civilizations would broadcast their presence in one way or another. This could take the form of a

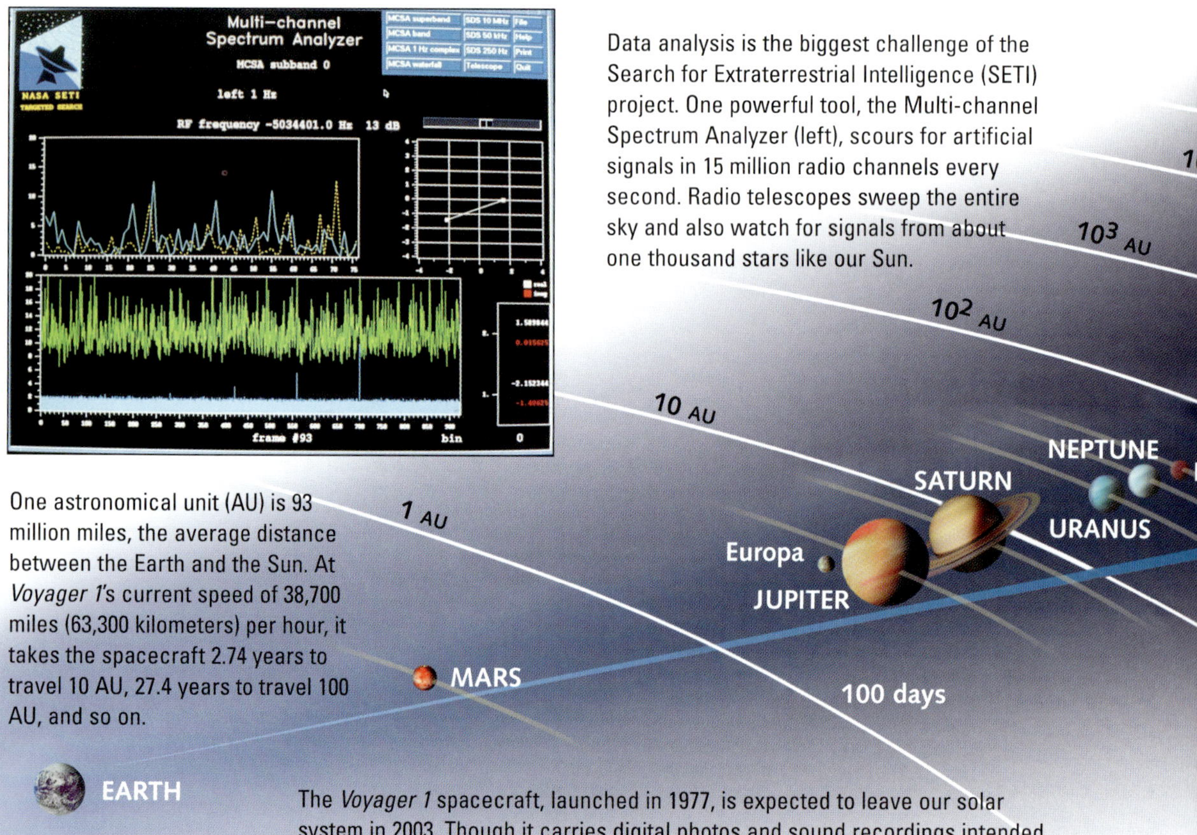

Data analysis is the biggest challenge of the Search for Extraterrestrial Intelligence (SETI) project. One powerful tool, the Multi-channel Spectrum Analyzer (left), scours for artificial signals in 15 million radio channels every second. Radio telescopes sweep the entire sky and also watch for signals from about one thousand stars like our Sun.

One astronomical unit (AU) is 93 million miles, the average distance between the Earth and the Sun. At *Voyager 1*'s current speed of 38,700 miles (63,300 kilometers) per hour, it takes the spacecraft 2.74 years to travel 10 AU, 27.4 years to travel 100 AU, and so on.

The *Voyager 1* spacecraft, launched in 1977, is expected to leave our solar system in 2003. Though it carries digital photos and sound recordings intended for extraterrestrials, *Voyager* would take 70,000 years to reach Proxima Centauri, the star nearest to our Sun. Sensitive radio telescopes on Earth are now listening for alien equivalents of "I Love Lucy" and other possible extraterrestrial signals from stars within 200 light-years.

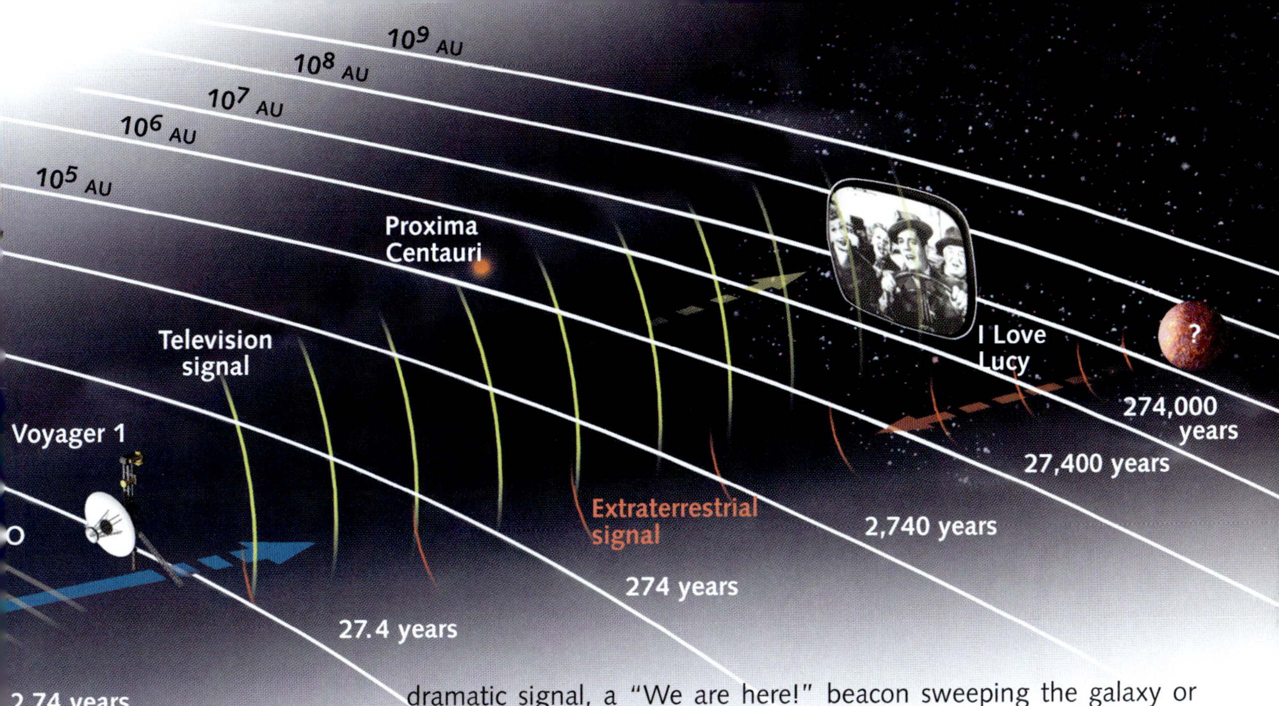

dramatic signal, a "We are here!" beacon sweeping the galaxy or targeted at nearby stars. Alternatively, a civilization's internal communications might leak unintentionally into space, just as our own radio and TV programs do. Routines by Burns and Allen and broadcasts of each Super Bowl are streaming outward at the speed of light within a growing bubble around our Sun. Advanced beings within a few dozen light-years could even now be decoding these faint signals, wondering what they signify about our young race.

In any event, SETI adherents believe that the choicest long-distance lines of communication lie within the radio bands. Radio waves travel virtually unimpeded across the galaxy, and they're easy to produce. Much of the radio spectrum is uncluttered with noise from astronomical objects. The flip side is that there are billions of radio channels to monitor. No one knows which channels are the likeliest ones, because there's no intergalactic programming guide. With so many channels—and with so many stars that populated planets might orbit around—SETI is the needle-in-a-haystack problem on the grandest scale.

When astronomer Frank Drake conducted the first SETI search in 1960 from an 85-foot radio telescope in Green Bank, West Virginia, he scanned two nearby stars at a handful of radio frequencies. By the turn of the century, SETI's equipment had grown a hundred trillion times more powerful. The most ambitious program, called Project Phoenix, is run by the SETI Institute in Mountain View, California. Project Phoenix astronomers scour for signals from a thousand Sun-like stars—all within 200 light-years of Earth—at 2 billion frequencies. The project uses a few weeks of observations each year on the world's biggest radio telescopes. Future plans call for an array of hundreds of smaller telescopes that will listen full-time for artificial signals from stars. Ultimately, SETI proponents hope to place an observatory in the quietest setting of all: the far side of the Moon.

A future array of radio telescopes might scan the heavens for extraterrestrial signals from a quiet orbit around Earth. By then, our civilization may have colonized the Moon (right).

The Arecibo Observatory radio telescope in Puerto Rico, the world's largest, is used for both astronomical studies and the Search for Extraterrestrial Intelligence. Nearly forty thousand aluminum panels reflect radio waves to a receiver overhead.

How will SETI astronomers know when they have detected a real message? Interference from sources on or near Earth is a problem. The signal must come from a distant point in the sky, and it must be repeated, preferably at different times of the year. Further, the radio waves must be artificial rather than natural. That's a surprisingly tough standard, since some astronomical objects can churn out radio signals with eerie regularity—including the rapidly spinning dead stars called pulsars. The content of a real message could be anything: a speech, an encyclopedia, or (as in the movie *Contact*) an instruction manual for building an interstellar transporter. Decoding the message may be the toughest nut of all to crack.

A parallel approach to SETI is to use visible light as a search tool. Stars are bright sources of light, but they shine vigorously in all colors of the optical spectrum. A civilization could transmit information by concentrating energy into a narrow part of the spectrum—a particular wavelength of red light, for instance. It's possible in that way for an artificial beacon to outshine a star at a certain wavelength. The effect would be the same as seeing a laser pointer across a crowded stadium. The lights of the stadium are bright, but the pointer focuses its energy into one narrow beam of sparkling color that the eye can detect. As more planetary systems are discovered, astronomers plan

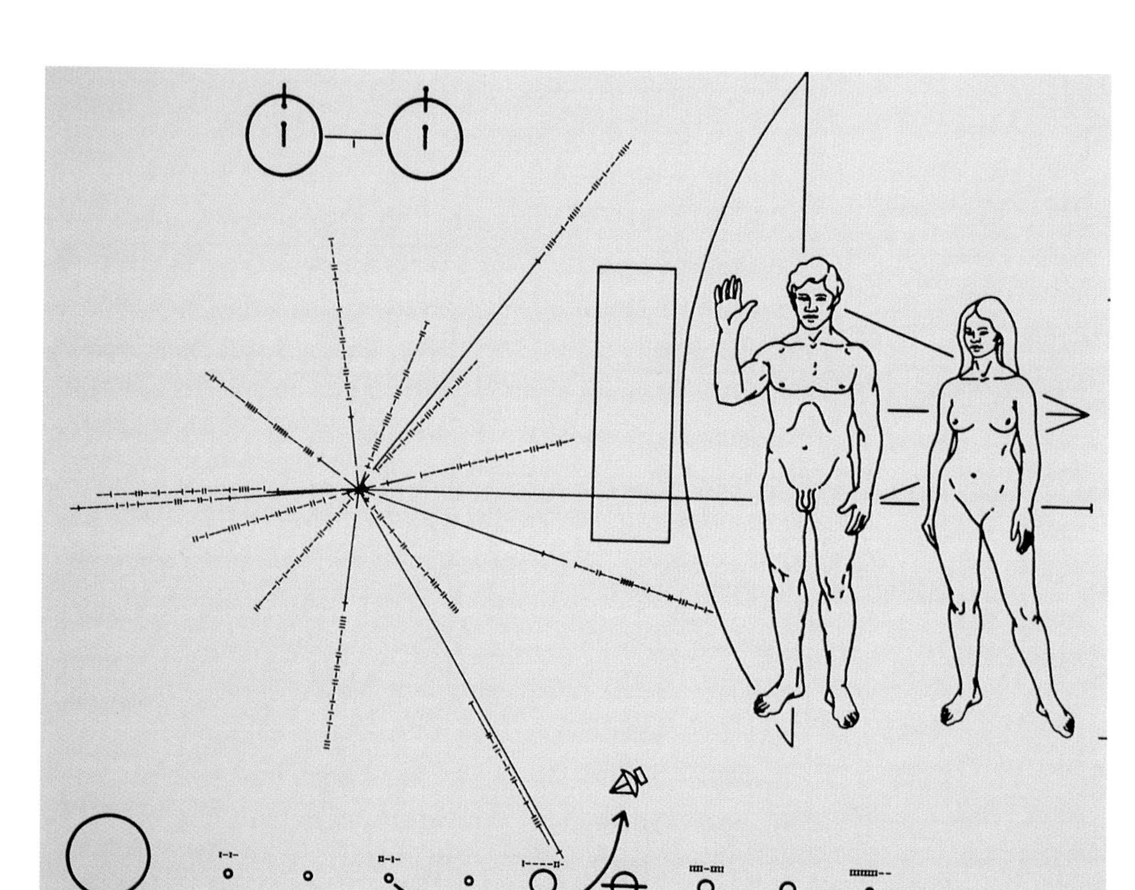

Astronomers Carl Sagan, Frank Drake, and others designed this plaque for the *Pioneer 10* and *11* spacecraft as a symbolic greeting from our civilization.

to scan them for such brief flashes. Intense pulses of laser energy that last only a billionth of a second also might do the trick if focused into a sharp beam by a big telescope mirror.

These studies clearly put Earth into a passive "receiving" mode when it comes to SETI. There is no consensus in our society to broadcast information about ourselves into the sky, other than the continual radio and TV leakage that already occurs. Nor is it clear that we would want to do so, given the potential hazards of alerting belligerent civilizations to our presence and our weaknesses. Astronomers have sent intentional greetings on a few occasions, but they were largely symbolic. For instance, the Arecibo Observatory radio telescope in Puerto Rico sent a three-minute radio signal in 1974. It contained coded information about our planet and our biology. The signal is streaming toward M13, a star cluster about 25,000 light-years away in the constellation Hercules. If anyone replies, we won't know for fifty thousand years. The *Pioneer* and *Voyager* spacecraft carry plaques and gold-plated audio recordings respectively, with descriptions of Earth and its people. These will drift slowly out of the solar system and into the vast expanse of the galaxy.

Many people dismiss SETI on philosophical or religious grounds, while others equate SETI with pseudo-science or UFOs. But in any serious discussion of SETI, it's useful to invoke something called the Copernican principle. Five centuries ago, Polish astronomer Nicolaus Copernicus held that there is nothing special about Earth's place in the cosmos or the time in which we live. We inhabit an ordinary planet that orbits an average star. According to this view, those physical conditions must exist elsewhere, and we are not the Milky Way's sole denizens. However, until SETI researchers find a solid signal, that argument will remain a rational chain of logic and no more.

Ultimately, SETI says just as much about our own species as it does about other life that might populate the cosmos. Humans are so curious about the natural world that we constantly push frontiers, whether to the depths of an ocean trench or to the surface of a Jovian moon. Searching for kindred organisms in the galaxy is the natural extension of that insatiable curiosity. The question of whether we are alone has shadowed our species since our first glimmers of self-awareness and our first sentient glances skyward. Today, technology finally has given us the tools to listen rather than just wonder. It would be foolish indeed not to try.

The two *Voyager* spacecraft each carry a gold-plated phonograph record with sounds and digitally encoded photographs from Earth.

ABOUT THE AUTHOR

Robert Irion is a freelance science journalist in Santa Cruz, California. He studied earth and planetary sciences at the Massachusetts Institute of Technology and science journalism at the University of California, Santa Cruz, where he now teaches a course in science writing for magazines. Irion is coauthor of *One Universe: At Home in the Cosmos* (Joseph Henry Press, 2000), a coffee-table book that describes the unity of physical laws on Earth and throughout space. He is a contributing editor at *Astronomy* and a contributing correspondent at *Science* magazine. He also has written for *New Scientist*, *Highlights for Children*, and other publications.

EXTRATERRESTRIAL LIFE

Picture credits

Cover Julian Baum/Science Photo Library/Photo Researchers **Back cover** NASA/Epix/Liaison Agency **Gatefold** Christopher Short **Endsheets** John Sanford/Starhome Observatory **Title page** Roger Ressmeyer/Corbis **5** David Nunuk/Science Photo Library/Photo Researchers **7** David A. Wagner/Phototake **8** Rainon/Explorer/Photo Researchers **11** A.B. Dowsett/Science Photo Library/Photo Researchers **12** Jean Claude Revy /Phototake **15** Alfred Pasieka/Science Photo Library/Photo Researchers **16-17** Tony & Daphne Hallas /Science Photo Library/Photo Researchers **19** Luis Garrido/Liaison International **20** Jon Wilson/Science Photo Library/Photo Researchers **22-23** Mauritius, Gmbh/Phototake **22** (inset) Jozon Michael & Agency HOA QUI/Phototake **23** (t) Laurence Pringle/Photo Researchers **23** (c) Sinclair Stammers/Science Photo Library/Photo researchers **23** (b) John M. Burnley/Photo Researchers **25** Phil Farnes/Photo Researchers **26** B. Murton/Southampton Oceanography Centre/Science Photo Library/Photo Researchers **29** Phillip Hayson/Photo Researchers **30** Mark Newman/Phototake **32-33** NASA/Science Photo Library/Photo Researchers **35** Daniel J. Cox / Liaison International **36-37** Tim Davis/Photo Researchers **39** (inset) Lowell Georgia/Corbis **40** Associated Press/NASA **41** NASA/Epix/Liaison Agency **42** NASA/JPL/Photo Researchers **42** (inset) NASA/Science Photo Library/Photo Researchers **45** NRSC LTD/Science Photo Library/Photo Researchers **46** NASA/Associated Press **49** Lynette Cook/Science Photo Library/Photo Researchers **50-51** Magrath Photography/Science Photo Library/Photo Researchers **53** Luke Dodd/Science Photo Library/Photo Researchers **56** SETI Institute/Science Photo Library/Photo Researchers **56-57** Don Foley/NGS Image Collection **57** (inset) Photofest **59** Dr. Seth Shostak/Science Photo Library/Photo Researchers **60** David Parker/Science Photo Library/Photo Researchers **62** NASA/Photo Researchers **64** Corbis